Ronaldo Batista Teófilo

Green Energy

Ronaldo Batista Teófilo

Green Energy

The Potential of Fruit Waste in the Brazilian Semi-Arid Region

ScienciaScripts

Imprint

Any brand names and product names mentioned in this book are subject to trademark, brand or patent protection and are trademarks or registered trademarks of their respective holders. The use of brand names, product names, common names, trade names, product descriptions etc. even without a particular marking in this work is in no way to be construed to mean that such names may be regarded as unrestricted in respect of trademark and brand protection legislation and could thus be used by anyone.

Cover image: www.ingimage.com

This book is a translation from the original published under ISBN 978-620-6-76133-4.

Publisher:
Sciencia Scripts
is a trademark of
Dodo Books Indian Ocean Ltd. and OmniScriptum S.R.L publishing group

120 High Road, East Finchley, London, N2 9ED, United Kingdom
Str. Armeneasca 28/1, office 1, Chisinau MD-2012, Republic of Moldova, Europe
Printed at: see last page
ISBN: 978-620-7-95850-4

THANKS

I would like to express my gratitude to everyone who, directly or indirectly, contributed to the creation of this book.

In first place, one acknowledgment The my mother, my brothers and my sisters in Recife – Pernambuco, people with whom we shared many good moments to concentrate our efforts on carrying out this work and who, despite the distance, always motivated me and gave me strength and courage to continue.

Register O my respect It is admiration at person of Prof. Dr. Manuel Rangel, It is Teacher Dr Clecia Pacheco what ever they were people, what They gave me all the assistance with any queries I had during the work, I would like to thank you very much.

To my wife, Joseilda, and my daughter, Tereza, who were in this endeavor with me, facing the day-to-day obstacles, Davi who was present contributing with doubts in a foreign language.

At the however, The implementation of this work no he would be possible without the help and total availability of my advisor, Professor José Lincoln Araüjo, and my co-supervisor, Professor Manuel Rangel Borges Neto. To both of you, my thanks and gratitude.

Thank you to Professors Antônio Carlos, Regina Aguiar, Claudemiro, Helena Paula and Washington Soares, my sincere thanks throughout the master's study journey and much gratitude.

No it could to leave in manifest The admiration It is respect what I have my work friends, teachers, who contributed a lot to this endeavor, Ana Maria, Sérgio de Carvalho Paes Andrade, Clésio Silva, Luis Carlos, Hommel Almeida, Ricardo Maia and Radmila Machado, gratitude to all of you.

"O bigger enemy of Knowledge is not ignorance, but the illusion of knowledge."

(Stephen Hawking)

SUMMARY

1. INTRODUCTION

O Brazil it is in between you 10 pafses what more waste food in the world. Around 35% of all agricultural production goes to waste. This means that more than 10 million tons of food could be at table of the 54 millions in Brazilians what live below from the line of poverty. According to data from the Social Service of Commerce (Sesc), R$ 12 billion in foods they are played outside daily, one sufficient quantity for to guarantee coffee from the morning, lunch and dinner for 39 millions of people (CARVALHO, 2009).

Currently, agro-industries have increasingly invested in processing capacity, generating enormous amounts of by-products, some of which are reused as animal feed.

Still, in many cases you by-products they are considered operational cost for companies, thus a large quantity is discarded and acts as a source of contamination. It is estimated that processing in fruits for production in juices It is pulp generates in between 30 and 40% agro-industrial waste.

A production of these by-products generates thousands in tons, It is important to add value and economic interest to these wastes, which require in one investigation scientific It is technological for enable their use since they are rich in bioactive compounds.

The waste generated by agro-industries varies, as it depends on the type of fruit to be processed, and this by-product may be composed of peels, seeds, seeds and even the pulp.

These residues are composed of vitamins, minerals, fibers, compounds antioxidants It is nutrients essential for O good functioning of the human organism; however, they are wasted in most factories and industries. It is reality It is worrisome visa what they he has big potential for a rich new food source, and minimize food waste. Several studies demonstrate the presence of important nutrients such as vitamin C, phenolic compounds and carotenoids in fruits, with higher concentrations in the seeds and skins. The intake regular in foods what he has these compounds it is associated with beneficial effects on human health.

Due to these high rates of production of agro-industrial waste, which, as indicated, are rich in nutrients and compounds essential for the proper functioning of the organism human, provide the motivation to carry out this research, with the purpose of preparing an inventory of organic residues available from fruit processing, their potential for energy use, us Limits of the municipalities in Petrolina – FOOT It is Juazeiro - B.A.

The municipality of Petrolina, located in western Pernambuco, in the of semiarid, It is one example in change, through from the insertion of technology. Until the mid-1980s, the products grown in the region were mainly short-cycle crops (tomatoes, watermelons, beans, onions, melons, etc.). After this date, a process of replacement and specialization in the cultivation of perennial fruit trees (mango, grape, banana, coconut, guava, among others) began, expanding the process of spreading scientific agriculture and agribusiness in the region. Between 1995-1998, the area cultivated with short-cycle crops (corn, beans, cassava, tomatoes, among others) shrank by 15%, while that of perennial fruit plants (grapes, mango, coconut, guava in between others) grown up 51% (SANT A-N-A, 2011).

In this sense, the circular economy is a strategic concept that is based at reduction, reuse, recovery It is recycling in materials and energy. Replacing O concept in end in life from the economy linear, for new circular flows of reuse, restoration and renewal.
In an integrated process, the circular economy is seen as a key element to promote the decoupling between economic growth and the increase in resource consumption, a relationship hitherto seen as inexorable.
Therefore, energy efficiency is a topic whose objective is to optimize processes to avoid wasting energy, generating savings, reducing costs and mainly contributing to the preservation of the environment, therefore, the industry perceives the need to obtain in energy through in sources renewable, although, with one good cost-benefit ratio. However, in Brazil and around the world, the use of biomass as an energy source is being increased every year, seeking to encourage the valorization of environmental, social and economic aspects, therefore, residues or organic by-products from various processing can be recovered in the form of renewable energy, which can be used in the same place where it was generated. Thus minimizing logistics and creating possibilities for energy self-sufficiency for various products. In this way, it was proposed to investigate, within the limits of the municipalities of Petrolina- FOOT Juazeiro- BA, to the companies in processing in fruits in operation, identify the ways in which organic waste is disposed of, evaluate the possibility of reuse in the process in the form of thermal, electrical energy, or even another sustainable destination identified at search In possession of the results, representa-los na form de infographic to leave de hardware stores de Geographic Information System.

2. OBJECTIVES

2.1 Geral

To understand as if from the The production of the waste organic available of processing in fruits us Limits of the municipalities of Petrolina – PE and Juazeiro – BA, and the potential for energy use.

2.2 Specific

• Iidentify It is to rank to the companies generators of waste;

• To analyze The availability in organic waste in companies that process fruits. ;

• Iidentify shapes in destination in your waste. ;

• To assess The possibility of utilization energetic of organic waste.

3. REVISION IN LITERATURE

One of the main obstacles in the development of the minimal fruit processing industry is the fact of generating a large quantity in waste organic, you which lots of times, no have a specific destination, becoming environmental contaminants. As the amount of waste generated can reach many tons, adding value to this by-product is of economic and environmental interest, with the need for scientific and technological research that enables its efficient, economical and safe use (NASCIMENTO et al . , 2015).

3.1 Utilization energetic

The production of renewable energy is one of the ways of using in waste organic at the Brazil, contributing to the at the same time to increase the global energy matrix, reduce launches inappropriate at the quite environment, generation in business opportunities, income and bringing greater sustainable development. Among the possibilities for energy use of organic waste are the production of biogas and biofertilizer, biodiesel and the production of briquettes or pellets (NASCIMENTO et al, 2015).

3.2 Biogas

"The production of biogas comes from the creation of a suitable environment for what to the bacteria methanogenic act about The matter organic and produce that fuel The leave in one route biological defined. This environment favorable The production of biogas refer if to the conditions chemical and physical conditions necessary for the development of these bacteria within the biodigester, which are determined in certain ranges of temperature, pH and carbon/nitrogen ratio (C/N) of the biomes" (ARAUJO, 2002)
Biogas is obtained from the decomposition of organic matter in reactors under anaerobic conditions and controlled temperature, which can be obtained in three main routes, the use in sanitary landfills (urban waste), industrial effluent residues or sewage (ETE) and waste or material originating from livestock and agriculture. Essentially these are the results of this process: one better control in passive environmental, The

possibility from the use digestate as a biofertilizer and the production of biogas for energy purposes, whether in thermal processes or electrical energy generation (BORGES NETO and CARVALHO, 2012).

According to Borges Neto and Carvalho (2012), it is important to assess energy demand in the region, considering the needs of rural properties, agro-industries and local communities. It is possible that the production of biogas from organic waste is a viable alternative to meet part of this demand, as long as there is adequate infrastructure for its transportation and distribution.

Second Freitas et al. (2019), The technology in production in biogas is still incipient in the country and, for a long period of 40 years (1970 to 2010), it did not receive adequate importance, being considered a by-product without value economic. A leave in 2010, O biogas it happened The be used at generation in energy, being considered one active energy and, no more, one passive environmental. A composition of biogas produced from anaerobic digestion is directly related to the type of organic matter decomposed. According to research carried out by the same authors, in Brazil, The technology in production in biogas yet if finds in early stage, having been neglected during one sorry in four decades, stretching from 1970 to 2010, when it was considered a by-product of insignificant economic value. From 2010 onwards, there was The to recognize O biogas as source in energy, converting it into a valuable energy asset, rather than a passive environmental waste.

In 2005 The Law no. 11,097 introduced O biodiesel at headquarters Brazilian energy in a more evident and significant way for this production. The law it says respect The "increment in the bases economic, social and environmental, The participation of the biofuels at headquarters national energy" and establishes guidelines for extensive use of other energy sources sustainable (BRAZIL, 2005). O biodiesel per your turn, It is derived from in fats It is oils per means chemical or biochemistry. The process involved in its production starts from the oil, as the raw material goes through transesterification generally using sodium hydroxide or hydroxide in potassium as catalysts for training of this product It is if subdivide in phases until to reach O goal Final, biofuel (SHIMADA et al., 2002).

3.3 briquettes egg pe//ets

"Briquets and pellets are agro-energy products obtained from the compaction of biomass, which replace firewood both for use in homes and in industries and commercial establishments as potteries, bakeries, pizzerias, industries chemistry, textiles and in cement. They they can to be produced with waste in wood, rice, corn, coffee, cotton, sugar cane It is several others, "preventing these materials from being left for natural decomposition, without using the energy contained in them" (EMBRAPA, 2013).

You briquettes,or pellets, they are one form more refined from the biomass produced per compaction (densification) in matter organic of plant or animal origin .The use of lignocellulosic materials It is O more commonly used having as waste example in sawmill, wood shavings, bark in rice It is remains in pruning (PAULA, et al , 2011). O briquette replaces The firewood for burn direct with the advantage, due to homogeneity, of taking up less space; issue any less smoke per account from the low moisture, ease at the transport, It is provide one destiny useful to the waste organic (TAVARES, et al , 2011). According to the same author, the use of briquettes as substitutes from the firewood in Law Suit in burn direct it presents several advantages, such as the reduction of occupied space due to homogeneity, the reduction of smoke emissions due to low humidity, ease of transport and the possibility of giving organic waste a useful destination.The way communication satellite technology has evolved a lot with O to spend of the years, especially for the wide possibility of its use in the most diverse sectors and, also, its contribution to reducing borders between countries.

Within this process of evolution of satellites, remote sensing emerges, which leads us to understand how, in fact, O arrangement at construction in maps, leading O understanding of the dimension of geoprocessing, explaining with its tools what we call in software for what, in fact, occasion The construction of maps. In order to feed data for the software, which will store precise information regarding what has been produced in fruit processing in industries in Petrolina - PEe Juazeiro - B.A.

3.4 Systems in information geography (sig)

Georeferenced Information Systems or Geographic Information Systems (SIG), from English, Geographic Information System (GIS) are a set of technologies that allow the analysis of spatial data, rationalizing the understanding of the processes of occupation of physical space. Its main tools are Digital Image Processing, Geostatistics and Georeferencing of data based on a coordinate system, most commonly, UTM (Universal Transverse Mercator) coordinates (SILVA, 2003).

In agreement with Silva (2003) O system in information geography is technology fundamental for O management It is The analysis in geospatial data, contributing to a better understanding of geographic phenomena and making more informed decisions.

The increasing availability of SIG software and platforms, as well as the democratization of access to geospatial data, further expands the possibilities for applying this technology in different areas, such as urban planning, natural resource management and environmental monitoring. According to Moreira ,(2003), what characterizes a SIG is the ability to interact spatial information from different computational sources, developing algorithms that allow quick consultation, plotting on a graphical representation of space and the possibility of new inferences regarding understanding reality from reading these images. Silva, (2003), points what, basically, one SIG It is compound per one source of data (coming from one or more platforms) and that can respond to the user quickly, with easy interpretation. One SIG no it needs to meet The all those types in analysis, but at least The one from them. Other functions assigned to a SIG are: produce maps in manner more quickly It is cheap, to allow experiments with different representations geographical areas, speed up access processes, comparison It is quantification in phenomena space (SLVA, 2003). Historical origin and application of Geographic Information System Second Golf (2011), O concept in analysis systematic of spatial data dates back to the 19th century. XIX, in the classic example of the English doctor John Snow what, in 1854, to the plot about one map in London the locals in occurrence in cholera during one phase in outbreak, managed to associate these positions with you habits in consumption in water of the population The one pit in where you infected collected water for to drink. By analyzing the water from this well, he was able to determine the vector causing the disease.To the to create one systematic in record It is catalog of occurrences, map them, select possible variables in relationship cause It is It is made, John Snow established to the bases of what if conventional to call subsequently from Systematic Analysis of Spatial Data (BOLFE, 2011).

A Systematic analysis of Spatial Data began along with the birth of Epidemiology and Medical Geography. She started from the principle that the understanding of different phenomena, as well as the understanding of their causes, can become much clearer when you data they are positioned about bases cartographic (SILVA, 2003). At the case mentioned, of doctor John Snow, if he If you only observen the addresses of those infected, but not positioning them on the map, it would take much longer or you might not be able to delimit O epicenter in where radiated O focus poisoning.

According to computer technologies and satellite astronavigation orbitals they were if developing to the far away of decades from 1960 to 1980, both fields in knowledge present intersection at Analysis Systematic in Data Spatial, giving rise to the expression System of Geographic information .Silva, (2003), points what the evolution of integrated Geographic Analysis to the world of data computerization and the modern conception of a GIS took place in 1962, when The company Tomlinson Group of Canada it happened to offer a service data storage in digital format, capable in accomplish operations in 19 overlap in polygons, create new areas polygonal about you maps It is synthesize you data in images in lists It is reports statistics in mode immediate to the user.You years 90 mark The diffusion in projects in SIG, incorporated by entities government It is big ones corporations business, in the purposes more diverse: mineral prospecting, environmental monitoring, real estate inspection, military surveillance, support for traffic systems, among others (BOLFE, 2011). As stated by Bolfe (2011), the 1990s are characterized by the dissemination of Geographic Information Systems (SLG) projects, which were adopted by government entities It is big ones companies in varied purposes, such such as mineral prospecting, environmental monitoring, real estate inspection, military surveillance, support for traffic systems, among others.In that same decade, a discussion that gained relevance regarding to the guidelines in construction of the SIGs he was at modeling cartographic representation of data, using matrix representations or vector representations.

In matrix representation, the fundamental unit is a regular polygon denominated pixel, where to the information from the area in pixel coverage are converted into a single value, based on the average incidence within that area. In the matrix model, data manipulation is much clearer, but its analysis is limited. This is the format of representation of images obtained per sensors orbitals, composing the first phases of geoprocessing development (MOREIRA, 2003).

According to the same author, the matrix model of data representation presents greater clarity in the manipulation of information, but its analysis is limited. This model is used to represent images obtained by orbital sensors and was widely used in the first phases in development of geoprocessing. In vector representation, data is positioned by a set of x and y coordinates, delimiting a point, line or geometric area with standard internal homogeneous. In general, that It is O final product of geoprocessing, when diverse spatial information of representations matrix they are synthesized It is grouped inside of these vector geometric shapes (MOREIRA, 2003).If at decade in 90 you SIGs represented highly complex technologies, whose clientele was restricted to government, academic scientific use or large business corporations, in the 21st century evolution of the gadgets in smartphones with resources integrated of Geopositioning, opened a window gigantic for the creation of SIGs in which data entry, its interface and the representation of responses are increasingly accessible to the lay public. In addition to orbital satellites, data collection stations or research centers, we will find, in the millions of cell phones spread with each user, a continuous and dispersed network of geopositioned information, collected voluntarily (LOTI, 2007).According to Loti (2007), in addition to orbital satellites and collection stations of data or centers of research, ha also a vast network of geopositioned information collected voluntarily by millions of users in phones cell phones scattered for the world. That network is continuous It is sprayed, O what means what there is one big amount of available geopositioned information. Large metropolis traffic monitoring systems currently operate with data provided by users of routing applications. Meteorological forecasting centers can increase the accuracy of their weather reports by using information provided by farmers in real time, via dynamic maps (WebMaps), while users can interact with the information presented in a much more conscious way (LOTI, 2007). As pointed out by Loti (2007), meteorological forecasting centers can improve the accuracy of their weather reports by obtaining real-time information provided by farmers, using dynamic maps (WebMaps). Furthermore, users can interact with the information presented in a more conscious way.The watchword of the new SIGs is "interaction", where platforms with different databases "talk" to each other, increasing their effectiveness.O client what access one application in delivery, without to perceive, This is passing data to another advertising application, which will search according to the customer's location and interests, which ads will be redirected to their smartphone.

It is with this characterization of these elements treated by the SIG in mind that we started to use this tool, as the main element of information regarding fruit productivity needs and their disposal, which are often wasted, O your volume in productivity, arriving to the point in or cause impacts to my environment.Being like this O SIG us leads The delimit It is bring about one arrangement in which it can show the use of these organic wastes, served in other physiographic areas for their legitimate use, whether in the form of energy generation or even as an element of fertility for the soil in the local environment.

With the greater processing power of current computing systems, and the possibility for much of this data to be stored at cloud, The discussion about representation vector or matrix if become any less relevant, with The availability in applications that can fully work, convert and manipulate these two representation models. Silva, (2003), states that the issues under debate in the new SIGs become other, such as thematic levels, the philosophy of the database oriented to the object, models hierarchical, modeling in 3D or incorporation of Artificial Intelligence into the system.

3.5 Structure in one sig

You SIGs work basically The leave from the interconnection harmonic five components: the working interface, the data input system, The Query It is analysis in data, The visualization It is plot It is the Geographic Database (MOREIRA, 2003).

According to Moreira (2003), Geographic Information Systems (SIGs) operate per quite from the integration harmonious in five components main: The interface in work, O system in Prohibited of data, querying and analyzing data, visualization and plotting and the Geographic Database.

Second O same author, those five components (Figure 1) are understood as: linterface: consists of a data integration base, allowed your streams of Bank in data to the system in analysis, thus forming the main bridge between the information that enters and that which leave of system. Prohibited in data: matches The all to the sources of data what feed O SIG, or it is, first stage in what The informations arrive to the system. AND compound by images in satellite, maps and images already processed by other SIGs and the tables where they are typed to the information. Query It is analysis in data: they are you software and applications that direct data entered into the input component to each other. Visualization and plotting: correspond to the tools that allow the system operator to see the SIG operations product, such as video monitors, touch screens and printers. Bank in Data Geographic: also called in base of data, are the information archiving systems, increasingly dispersed in the cloud. What distinguishes the Geographic Database of a SIG is that most of its information is linked to a spatial positioning or geographic addressing code.

Figure 1 - Scheme representative in components in one System Integrated Geographic

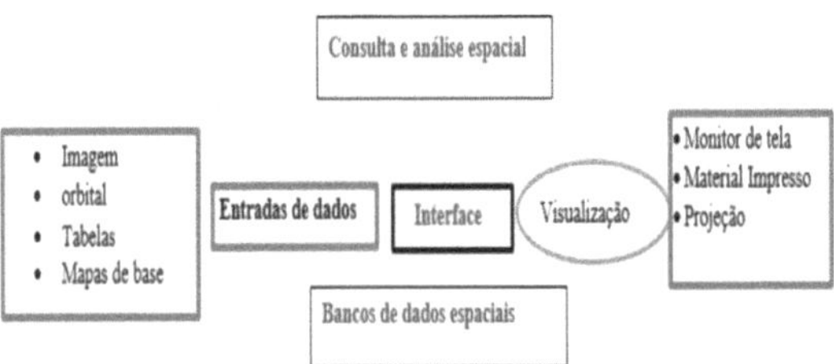

Source: Authorship own.

In general, due to the particularity of SIGs, which work with very large volumes of information, these banks can be divided in between many different agents of the sectors public It is private what are shared across the virtual network. Thus, an SIG can manipulate data that is under the responsibility in organs so distinct as IBGE, Embrapa It is Sebrae and yet, present uniform answers, giving a false impression that they all come from the same base. Ana/ise Systematic in Data Space It is Architecture of SIG. The subjectivity of the Systematic Analysis of Spatial Data, at all stages, should not be neglected, since both the sensor systems of orbital satellites and the digital images produced by processing applications are still attempts to quantify geographic elements, which lots of times They are not expressed numerically in the real world, but are products of abstractions in scales of values, created throughout human history (LUCHIARI, 2005).

Based on Luchiari (2005), the interpretation of the local environment, when if treats of arrangement space of the producers in fruits from the region irrigated of OK of They are Francisco, matches The one landscape what aims at Show one reality different of introduced in terms of utilization of the resfdous organic distributed at region, quoting yours interpretation cultural, while O what if realize It is The search of the economic productivity of the studied area. The selection between the countless possibilities for quantifying geographic phenomena, as well as in the elaboration of synthesis algorithms, It depends in one good dose in experience It is in knowledge of the space being approached, by the team of professionals involved in the elaboration of an SIG. As data undergoes statistical processing, this subjectivity in categorization, classification or separation can be amplified It is to generate big ones distortions us results finals, O what requires continuous verification and feedback of data based on new analysis (SILVA, 2003). According to Silva (2003), as data is subjected to statistical processing, subjectivity in categorization, classification or separation can be amplified, which can generate large distortions us results finals. Per that, It is necessary carry out continuous verification and feedback of data based on new analyzes to guarantee the quality and reliability of the results obtained.

This reflection is important because it shows the need for the architecture of the Geographic Information System not to be too closed into hierarchical levels of classification, as demonstrated in Figure 2, in order to prevent data alteration analgesics in between to the layers in processing, fact that what leads us to understand how to generate accurate data on fruit production and how to reuse organic waste, being a factor in interpreting and analyzing what has been causing changes in behavior of producer in fruits It is reuse of the waste organic, produced and carried out within the cities of Petrolina – PE and Juazeiro – BA.

Figure two: description of space physical /the CA/

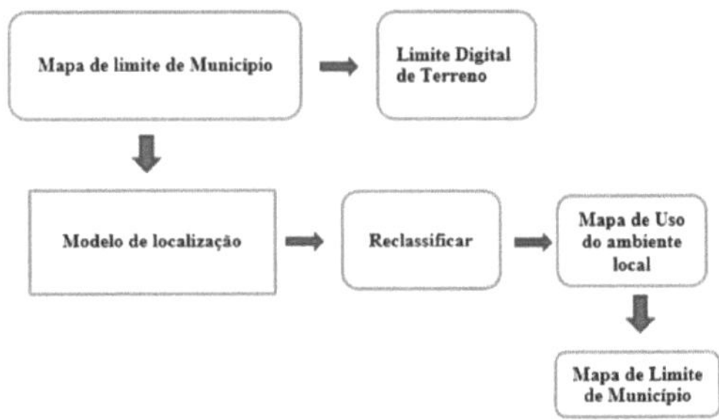

Source: Authorship own.

Since it is necessary to analyze and interpret the reuse of fruits at the environment from the industry in fruit processing from both municipalities, highlight the creation of SIG layers, which will be used for the analysis and creation of the inventory of organic waste available from fruit processing and their potential for energy use, in addition to this situation, we can also add a layer, which will show the fruit processing industries in the regions, taking into account the transport routes, ensuring that this layer will show the transport routes that interconnect to the areas in cultivation It is to the industries in processing in fruits. As well as analyzing the data with the SIG layers created, we can analyze the data to identify the organic waste available from fruit processing and its potential for energy use, to include identification of the types of fruit cultivated in the areas selected and identify to the quantities in fruits produced and processed in industries, in addition to this situation exposed, identify the types of organic waste generated during fruit processing. Another important characterization to mention is the quantification of available organic waste, which is consistent with the identification of transport routes that connect cultivation areas and industries. in processing in fruits. Like this we can with The using SIG tools, prepare an inventory of organic waste available from fruit processing and their potential for use.

3.6 Comp/exity of the studies environmental

In one quite field in between to the Sciences Social It is to the Sciences Exact, Life Sciences (such as Biology, Ecology and Agronomy) suffer one expressive nfvel in relativity in function of big number of variables what encompass, depending on in conditions space and very specific temporalities, where a procedure may be effective in a given situation and totally inappropriate in another, even when apparently similar.

In that sense, it was observed in some cases, what O owner of the farm producing organic waste, is unaware of the potential of organic waste inserted within its own property to be reused The production of the elements had as discarded for reuse of it , as the energy generating source.

That complexity of Sciences Natural open controversies well-founded It is needed at the scope academic, but also, give margin The visions partial, connected The dogmas ideological, economic interests or politicians, what invariably if enjoy of these Controversies, whether intentional or not, end up hindering the understanding of anthropogenic intervention strategies, aimed at exploiting the productivity of natural resources (GRAZIANO; GAZZONI; PEDROSO, 2020). The complexity in understanding that this is a tool that enables the understanding of what is produced by the information that will generate standardized SIG data is perceived in the authors' approach.

This controversy highlights how difficult it is to establish a methodology for analyzing environmental impact, even based on statistical data, without allowing one to lean towards one or another "preferred" method. in if to assess The problematic. With this it appears that in addition to food production, coming from organic materials from industrial in processing in fruits at delimitation of the municipalities of Petrolina – PE and Juazeiro -BA, you can see how much you have to reuse you waste organic, as offer in inputs at agricultural production, as an alternative source of energy generation in the search for environmental control and even a generator of sustainability for future generations, therefore there is an urgent need to delimit these producing areas through the creation of maps, as an identifier of industries in processing in fruits us two municipalities in largest volume in the São Francisco Valley.

3.7 Mapping of the waste organic

The mapping was produced with the volume of agricultural production and organic waste summarized in a spreadsheet, using Microsoft software *Excel* It is evaluated per Micro Environments physiographic of Municipality of Petrolina – PE and Juazeiro - BA. Taking into account that the resource seeks to guarantee the full functioning of activities in certain areas in one industry in processing in fruits. This makes it possible to identify errors and points for improvement to optimize resources.

Each businessperson local go to know The your function within the production chain, in addition from the goal of your work It is The importance the same.

For management, the use of the Geographic Processing System, to determine the mapping, leads to providing a macro view of the production processes, in which it is possible to see each stage in work for, of that form, identify points improving and optimizing routines.
This has a positive consequence on issues of waste in material organic from the company, visa what you waste are eliminated It is controlled from the routine in work being drastically r educed It is, per consequence, The performance in each area producer of organic waste is considerably optimized.

In addition of the points what detailed above, bring about The location and determine O map makes it possible The to guarantee O what if he does in utilization of waste organic, come bring about The routine in work, leading functionality in an optimized way, without waste or risks for the company and employees. Thus, it is possible to create a very detailed flowchart of each process industrial, in what O goal It is identify points in optimization or eventual failures what compromise you results business. Furthermore, map also makes it possible what one routine in Enhancement continuous it is established. Of that form, you Law Suit they are validated recurrently and, obviously, can evolve consistently. From the same form what eliminate failures It is Act us points in improvement, mapping industries also provides a great evolution regarding the rational use in the production of organic waste in their productivity. Thus, the company can improve your results and become more competitive in your segment in acting, leading a visualization from an environmental point of view, providing rational use and leading to sustainability and sustainable development.

According to EMBRAPA 2016 (Macrologistics of Brazilian Agricultural Production), the main products of the irrigated fruit growing sector in the São Francisco Valley are mango, grapes, sugar cane, wood (pine and eucalyptus), corn, acerola in addition to poultry, chickens, goat farming and general subsistence farming such as beans. Analyzing the production of the State of Pernambuco in this site, It is possible come to the sources primary in data collected by ABRAFRUTAS, according to Table 1.

Tabela 1: Fonte primaria de dados sobre produçao de frutas em Pernambuco.

Comparativo da exportação de frutas

Período: Os 11 primeiros meses de 2020 - 2021

Mês/Ano	Janeiro a Novembro de 2020		Janeiro a Novembro de 2021		Variação	
Frutas	Valor(US$)	Peso (Kg)	Valor(US$)	Peso (Kg)	Valor(US$)	Peso (Kg)
MANGAS	$ 214.450.875	210.664.868	$ 223.435.674	244.840.616	4%	16%
MELÕES	$ 119.188.218	192.572.054	$ 137.332.760	213.784.796	15%	11%
UVAS	$ 97.267.168	43.739.052	$ 137.163.956	67.267.295	41%	54%
LIMÕES E LIMAS	$ 95.245.337	110.718.295	$ 113.449.600	132.510.437	19%	20%
CONSERVAS E PREPARAÇÕES DE FRUTAS (EXCL. SUCOS	$ 64.078.539	44.553.315	$ 88.826.678	50.398.657	39%	13%
MAÇÃS	$ 41.239.430	62.551.774	$ 73.841.733	99.032.505	79%	58%
MAMÕES (PAPAIA)	$ 37.959.056	39.276.351	$ 46.142.894	46.082.673	22%	17%
MELANCIAS	$ 36.460.900	91.149.245	$ 44.775.313	101.182.004	23%	11%
BANANAS	$ 23.947.307	79.508.590	$ 33.816.424	100.447.027	41%	26%
OUTRAS FRUTAS	$ 17.892.622	8.407.547	$ 20.139.772	8.713.809	13%	4%
ABACATES	$ 13.069.225	7.491.552	$ 14.885.131	8.505.610	14%	14%
FIGOS	$ 3.539.884	884.558	$ 4.428.465	1.180.269	25%	33%
ABACAXIS	$ 1.797.562	3.014.475	$ 2.735.155	3.910.200	52%	30%
PÊSSEGOS	$ 811.267	704.538	$ 2.324.468	2.164.098	187%	207%
COCOS	$ 910.313	992.777	$ 1.114.064	915.915	22%	-8%
CAQUIS	$ 239.344	137.591	$ 1.067.057	912.312	346%	563%
LARANJAS	$ 4.245.525	6.853.068	$ 932.177	3.532.574	-78%	-48%
GOIABAS	$ 440.035	197.722	$ 899.067	407.140	104%	106%
MANGOSTOES	$ 2.840	1.922	$ 352.235	56.557	12303%	2843%
TANGERINAS, MANDARINAS E SATOSUMAS	$ 228.543	232.231	$ 219.431	201.303	-4%	-13%
PÊRAS	$ 182.429	84.178	$ 156.748	70.392	-14%	-16%
MORANGOS	$ 218.482	84.565	$ 143.212	39.854	-34%	-53%
KIWIS	$ 135.645	40.355	$ 114.796	37.894	-18%	-6%
CEREJAS	$ 109.513	12.413	$ 63.891	8.503	-42%	-81%
TAMARAS	$ 93.637	28.423	$ 48.021	12.257	-48%	-57%
POMELOS	$ 35.489	12.007	$ 24.159	8.080	-32%	-33%
AMEIXAS	$ 17.619	4.530	$ 14.866	2.998	-18%	-34%
DAMASCOS	$ 7.261	895	$ 5.151	613	-29%	-82%
MARMELOS	$ -	0	$ 304	144	100%	100%
TOTAL	$ 773.814.065	903.918.691	$ 948.453.202	1.086.228.532	23%	20%
FONTE: MAPA - AGROSTAT/MAPA	Valor(US$)	Peso (Kg)	Valor(US$)	Peso (Kg)	Valor(US$)	Peso (Kg)
ELABORAÇÃO: ABRAFRUTAS	Janeiro a Novembro de 2020		Janeiro a Novembro de 2021		Variação	

Source : Openfruits 2020/2021

The Brazilian Northeast, especially the semi-arid region, presents severe edaphoclimatic conditions that make intensive livestock farming such as cattle and pig farming difficult, with sheep and goat farming predominating.The region is also home to the Caatinga biome, exclusive to the Northeast region. of Brazil. Your economy It is basically in livestock extensive and low-income family farming. However, this region is economically viable, as long as there is adaptation to the environment to

The coexistence with O semiarid, per quite of respect The nature It is The agroecological production with economic autonomy and harmony with the environment (SILVA, 2016).The National Solid Waste Policy (PNRS) — Law no. 12,305/2010 (BRASIL, 2010) prioritizes non-generation, followed by reduction, reuse, recycling, treatment of solid waste and environmentally appropriate final disposal of waste. According to this policy, reverse logistics and shared responsibility, established by law, are strategic in the implementation of a production model It is consumption sustainable.
To the responsibilities of generator, of importer, of distributor, of the merchant, like this as of consumer of product, with one reverse flow of waste, can positively impact this objective.

However, according to Campos (2012), this path has not yet been built and even in the European Community countries, with their strict directives, the results are not auspicious, which keeps the possibility of developing countries reaching better levels of achievement even further away. that we currently have (SILVA, 2016b).

One form in mitigate to the questions environmental It is costs involved in processing of fruits, It is O eventual utilization of the organic waste in the unit itself, which can be thermal processes, the production of electricity, or even by-products that can be sold.
An example is the sugar and alcohol industry, which uses the burning of sugarcane bagasse for thermal processes and electricity generation, even exporting the surplus energy produced. To the companies in processing in fruits, as to the packing house , and asqueindustrialize you products they are subject The production of waste organic It is emission in effluents. O service to the what determinesThe law, what for O greeting from the PNRS, ends per generate additional costs to the process, which can negatively impact the final price of the product and, consequently reduce competitiveness in the market One of the main obstacles in the development of the minimal fruit processing industry is the fact of generating a large quantity in waste organic, you which, lots of times, no have a specific fate, becoming environmental contaminants. As the amount of waste generated can reach many tons, adding value to this by-product is of economic and environmental interest,with the need for scientific and technological research that enables its efficient, economical and safe use.Waste agro-industrial they are treaties as by-products obtained from of processing industrial in foods. O treatment

of these remotes the decade of 1970 by reusing waste from the bark of certain fruits as raw material in order to produce food perfectly passable in be included at food human. As shown in Table 2, the nutritional composition of residues from various fruits.

Tabe two: Composition nourishes/ in waste in several fruits

Frutas	Residuos	Umidade %	Cinzas %	Proteína %	Lipídios %	Carboidratos totais%	Calorias (100 g)
Goiaba	Polpa	65,54±0,32	0,72±0,02	2,82±0,19	2,94±0,01	27,98	150
Goiaba	Casca	84,98±0,04	0,51±0,00	1,03±0,2	0,07±0,2	13	5
Acerola	Polpa	83,45±0,06	0,55±0,03	1,65±0,26	3,59±0,15	10,76	82
Abacaxi	Polpa	88,19±0,80	0,53±0,04	1,05±0,01	0,69±0,03	9,54	49
Graviola	Polpa	83,16±0,83	0,48±0,04	1,09±0,07	2,28±0,13	12,99	77
Bacuri	Polpa	83,81±0,13	0,65±0,28	0,56±0,03	3,84±0,02	9,14	74
Cupuaçu	Polpa	93,86±0,08	0,20±0,12	1,65±0,38	3,69±0,02	0,6	42
Maracujá	Sementes	6,89±0,14	1,47±0,09	12,57±0,52	28,12±0,75	13,19	*
Maracujá	Casca	89,08±0,00	0,92±0,00	1,07±0,00	0,70±0,00	8,23	*
Uva	Casca	89,75±0,06	1,85±0,01	1,95±0,2	0,11±0,09	13,6	53
Manga	Casca	78,70±0,45	0,99±0,05	1,24±0,11	0,18±0,01	12,89	64
Manga	Polpa	82,11±0,21	0,34±0,06	0,44±0,08	0,61±0,03	16,6	*
Tomate	Fruto	95,88±0,05	*	0,66±0,04	0,26±0,01	10,42	*
Banana	Casca	3,30±0,08	2,59±0,07	4,50±0,84	*	*	373
Pera	Casca	81,87±0,03	0,33±0,00	0,41±0,1	0,13±0,15	14	53
Laranja	Casca	76,55±0,02	1,04±0,01	1,00±0,015	0,26±0,1	11,7	46
Melancia	Casca	96,0±0,9	0,58±0,04	0,93±0,4	0,30±0,02	2,19	24,74
Maçã	Casca	82,14±0,06	0,03±0,00	0,32±0,10	0,3±0,25	13,6	56

Source: Rev. Virtual Quim. 1Vol 71 1No. 61 11968-19871

This form, proposes investigate, at the limit of municipality of Petrolina -PE fruit processing companies in operation identify which to the shapes in destination in your waste organic, to assess The possibility in reuse at the process in form in energy, thermal, electrical, or yet, other sustainable destination identified in the research. Once you have the results, represent them at form in infographic The leave in tools in System Geographical in information.

3.8 Challenges of project

Two major integrated themes constitute the main focus of this dissertation, big amount in waste generated in the industries in fruit processing associated with the scarcity of energy and materials (chemical compounds) of low impact and renewable origin. The State of Pernambuco and its 184 municipalities produced in 2013 The amount in 59,291 tons /day in waste organic, 99 % were collected. The data indicates growth of 4.9% in the total collected It is increase in 4.7 % at generation in waste organic in rural area in relation to the previous year. In 2013, around 23.6% of the state's organic waste (corresponding to 13,865 tons daily) was destined to the "dumps" It is landfills controlled, you which of point in environmental view, little if differentiate of the own dumps, then no have the set in systems necessary for protection of quite environment It is of environmental public health . (ABRELPE, 2014). Organic waste can be used to improve the physical and chemical conditions of the soil, as natural fertilizers. Therefore, the use of compost in "organic waste" at "formulation in substrates for the cultivation of plants in containers contributes significantly for The improvement substrate fertility", as the organic waste compound presents large amounts of nitrogenous compounds, (LIMA et al., 2010). Mamede (2013) in his dissertation describes the main means of utilization energetic for attend The demand growing in energy, as well as established standards and policies in force in the socioeconomic and environmental spheres. The separation of waste wasted in farming into its most crude components: organic waste, recyclable waste and waste in general, allows different routes for using the different fractions: digestion anaerobic, production in "fuels derivatives in waste" for digested waste and incineration of organic and non-organic waste according to The Figure 1. Second O report revolution energy in an evolution analysis, considering the year 2003 until 2050, found that there will be an increase in efficiency in energy systems as well as an evolution in the use of non-renewable energy (KREWITT et al., 2007).But there are still many obstacles to be overcome and studied. One of the questions to be better evaluated is whether energy should be concentrated or decentralized. Apart from other questions, the answer to this question must be linked to important supporting information, that is, its location.The most promising and consolidated spatial data processing tool is the Geographic Information System (SLG). This tool has the ability to analyze spatially distributed phenomena through the variables that govern them.

Much of the waste generated in fruit processing is discarded into the environment. Therefore, the use of these wastes could contribute positively to the reduction of environmental impacts, enabling the production of products with high added value.

3.9 Characterizing to the measures expected

When it comes to the production of organic waste that is wasted in the fruit processing industry in the municipality of Petrolina - PE, it is necessary to delimit these areas where the fruit processing industries are located so that it can be used efficiently. sustainable being wasted by the fruit processing industry. In beginning, they were made queries in miscellaneous bases in data. It is estimated what O consumption Brazilian in fruits processed pass of the 23 million in tons, generating one immense amount in waste that are discarded unduly in the environment, which is no different when if treats from the region in fruit growing irrigated of OK of They are Francisco at the municipality in Petrolina - FOOT.One of the ways found to solve this problem is to know how to identify what would be the ways of disposing of the organic waste caused by fruit processing industries, as well as evaluating the possibility of reuse from the perspective of production in energy, it is in form theme, electrical or same another path that generates a sustainable destination. Based on the above, there are numerous applications for the reuse of waste from the fruit processing industry, contributing as a promising alternative for innovation in industries, minimizing damage to the quite environment It is promoting health and well-being for the population of the producing region.Treating in waste solids, they are many you details that,if not observed, they can compromise all The chain productive- when the segment in question It is O industrial – or same the services, when they are you establishments in producers in fruits. Ideally, fruit processing industries in these municipalities cited avoid O waste of material organic products produced by them. This way there can be more control over their generation It is one can to work you indicators for to decrease – or same reset - The production in waste. A form of work to be carried out to address this entire problem in the generation of organic waste, promoted by the fruit industries, notably at region of sub-medium

OK of They are Francisco, including the municipality of Petrolina – PE, will be prepared using a SIG tool (Geographic Information System), resulting in an inventory of organic waste, as having great potential for energy use, generating an atlas identifying environments with this production potential, originating from industries of fruit processing.

4. MATERIALS AND METHODS

The methodology adopted in the research was exploratory, descriptive, with elements quantitative It is qualitative, what if based in the answers obtained by companies what process fruits at the OK of They are Francisco. After the bibliographical review, a search was carried out through the National Register of Legal Entities (CNPJ) of food processing companies with activities in the

municipality in Petrolina, with views delimit O size from the sample The be searched. According to the availability of access and location, the objective was to conduct visits to companies, for possible measurement, or same, collect in samples for to determine, if required in the laboratory, the energy potential of the waste. In a complementary way, he was collected information via quiz or information made available by companies on their official channels, by applying the questionnaires set out in Tables 3 and 4.

Tabe 3: Frame what i/ustra to the questions used at the work/ho.

Location	O what it is in (local)?
Condition	Where it is?
Trend	O what changed (or will change) in determined location?
Route	Which O best way?
Standard	Which The variable dominant in determined locality
Simulation	As it will be The occurrence in one event in determined location?
Modeling	What changes will happen to determined event.
Location?	

Source: adapted in Silva (2003).

Tabe 4: Questions applied.

QUESTIONING 01	Query about The classification in accord with O size from the company: GP(Large Size); EPP (Small Business); ME(Medium company); MP (Medium Size).
QUESTIONING 02	Iidentification of number in companies in agreement with The material produced.
QUESTIONING 03	Query how much The classification of the waste It is effluents at the production process
QUESTIONING 04	Query how much to the process productive in relationship The use or no heat demand in the process.
QUESTIONING 05	For to the companies what use heat in your process productive, consultation on the use of the energy source used.
QUESTIONING 06	Query referent to the methods in use in treatment of waste.
QUESTIONING 07	Query about of knowledge how much The economy Circular.
QUESTIONING 08	Query about The application in projects in efficiency energetic in its production process.
QUESTIONING 09	Query about The possibility in visits for eventual diagnosis of energy efficiency.
QUESTIONING 10	Query about O knowledge about in Software in waste.

Source: Authorship own.

With the results obtained, companies were classified according to: their location (rural/urban), their size, original raw material, final product, volume and destination of organic waste, and possible technology for the associated energy use.

4.1 Processing in data

Using of map in limit of municipality in Petrolina - FOOT It is of the Municipality in Juazeiro – BA, O model numeric in ground, both available for the Institute in lands, Cartography It is Geodesics – ITCG, the geographic location was mapped using geoprocessing techniques, with the help of SIG. A definition of classes of occupation of the local physical space of establishments of research interest was based on the methodology proposed by location ,using GPS, being what you maps in guidance of the land of the two locations were prepared as presented in form illustrative, as show O flowchart from the methodology for generating the database (Figure 2).

4.2 Description from the area

A study area, Municipality of Petrolina, located in the state of Pernambuco at the semiarid of North East Brazilian, known as Sub-middle of the Sao Francisco Valley and with the following coordinates: 9° 23' 39" S; 40° 30' 35"O; It is 380 m (Figure 3). O Municipality presents 4,561.874 km2 in area territorial, with one population estimated for O year in 2018 in 343,865 people, density demographic he is in 64.44 inhabitant km2, and PIB per capita of R$ 17,160.36 (2016), being considered the largest in micro-region and fifteenth (15th) largest in the state, according to data from IBGE (2018). The region's climate is type BSwh, according to Kôeppen's classification, corresponding to a hot and dry semi-arid climatic region, with very irregular rainfall. The index rain gauge Yearly average It is 571.5 mm, distributed in between you months of December The April. A temperature average Yearly It is in 26.4 °C, with minimum average in 20.6 °C It is maximum in 31.7 °C (LOPES et al., 2018), It is with insolation in 3,000 hours.year -1 (SOBEL; ORTEGA, 2010).

Figure 3: Map in /oca/izacao space/ from the area in study.

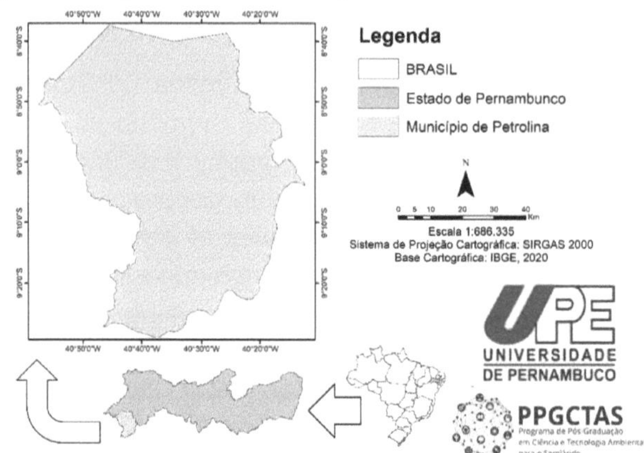

Source: Authorship own.

Given the information regarding the location of the municipality of Petrolina, the locations of the research reference points stand out. in question, where in home they were identified five grape and mango producing farms in the municipality of Petrolina, which were the project's references. Once the identification was carried out, it dealt with choices of where the color- laugh in the industries in processing in fruits, to the possible industrial properties that it established, the order in waste production organic at perspective in reuse of the waste in the transformation in generation in energy It is until same, O reuse of the plant in a productive form in the characterization of agricultural productivity.The environments in which their waste was reused as a way of complying with the principles of sustainability were outlined in order of productivity and generation of organic waste. So, through GPS, the geographical coordinates were determined, as an identifying factor of the productive areas and generators of organic waste that us made it possible in to create you maps themes. Nessa jor- anything in identification, geographical coordinates were read with the following fruit processing industries: Coopexvale, Frutirenda, Gdoces northeasterners, Timbauba Agricultural It is Vinfcula They are Braz, as shown on the map (Figure 4).

Figure 4: Map in images in sat/ites from the /hollow/age in study.

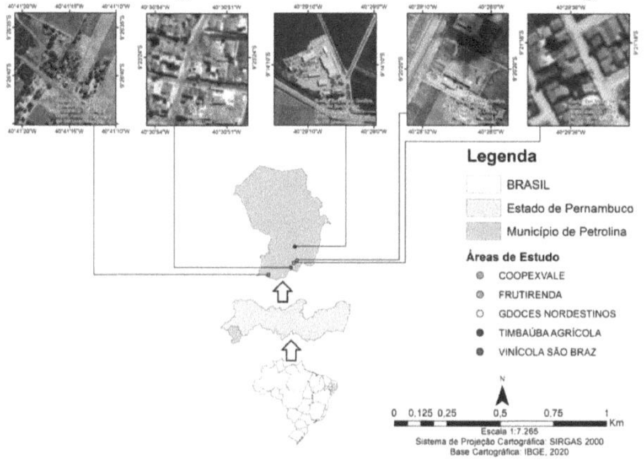

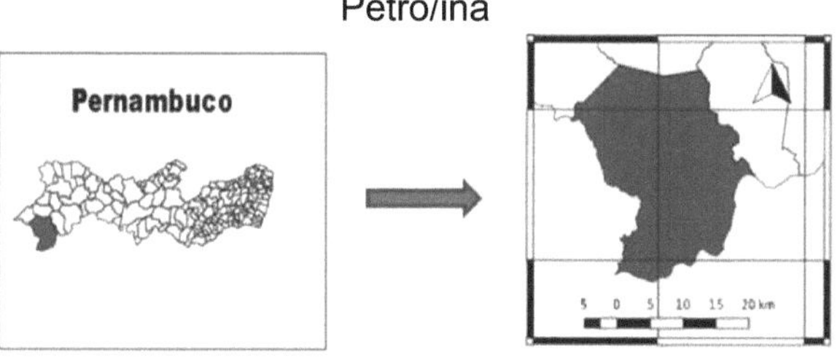

Source: Authorship own .

O municipality it has one area in 4 . 561.87 km2, bigger municipality in territorial extension in Pernambuco. Is situated The 09° 23' 39" in latitude south It is 40° 30' 35" longitude west of the state capital. The neighboring municipalities are: Dormentes to the north; state of Bahia (Juazeiro) to the south; Lagoa Grande to the east and, to the west, Afrânio and then again the state of Bahia (Casa Nova).IBGE (2018) The municipality is located in the geoenvironmental unit of Depressão Sertaneja, a unit that is formed by the main characteristics of the northeastern semi-arid region. Its relief is marked by a very monotonous pediplanation surface, being predominantly smooth-undulating and crossed by narrow valleys with dissected slopes (Figure 5).

Figure 5: Map in Pernambuco with emphasis to the municipality in Petro/ina

Source: Authorship own.

There are also residual elevations on the horizon line, ridges with/without hills. This type of relief bears witness to the intense cycles in erosion what reached O sertao northeast. A altitude The average for the municipality's headquarters district is 376 meters above sea level. Petrolina is inserted in the smaller microbasin of the Pontal River, inserted in the macrobasin hydrographic of River They are Francisco, in addition of group in Basins of Small Rivers linteriors. All your courses of water, with exception of São Francisco, are intermittent and have a dendritic drainage pattern. To the south of municipality if locate some of main islands of São Francisco, such as: Ilha do Fogo, Massangano and Rodeador (or Rodeadouro) (MOURA, et al., 2007).The main streams are: Baixa Salina, Pedra Preta, Baixa do Procôpio, Good Jesus, Earth New, from the Grotto Big, of Torch, Low Coveiro, Baixa do Boi, Estandarte, Formosa and Areia. The most important reservoirs in the municipality are: Vira Beiju (11,800,000 m3) , Salina (4 , 021 , 375 m3) Baixa do Icô (1 , 300 , 000 m3) and Barreira Happiness with capacity in two . 880.000 m3 in water. (CPRH, 2014) Yet if account with to the lagoons: from the Crafba, of Reed, from the Sand It is from Tapera. By superimposing a satellite image using the QGIS software, we initially determined the satellite image of Fazenda Frutirenda S/A and Fruticultura Alagoana LTDA, these two farms were the starting point of the work carried out with the Goog/e Earth imaging platform with QGIS engine expansion.O municipality in Juazeiro if extends per 6 . 500.7 km2 It is counted with 216 . 707 population at the last census. A density demographic It is in 33.3 inhabitants per km2 in the municipality's territory. Neighboring the municipalities of Petrolina they are Sobradinho It is Juazeiro what if situates The 5 km The South-East from Petrolina. Placed on The 369 meters in altitude, in Juazeiro he has to the following coordinates geographical: Latitude: 9th 26' 18" South, Longitude: 40° 30' 19" West.According toThe delimitation of project, treating yourself of the organic waste at industry in processing in fruits, delimited itself also the municipality of Juazeiro, as being an element of study in the dissertation. Your location geography if he does from the Following form: Juazeiro It is one municipality Brazilian of state from the Bahia, located in the Mesoregion of OK San Francisco from the Bahia It is Microregion in Juazeiro.Your population in 2019 it was in 219 . 544 population in agreement with the I estimated of IBGE (2019), being O fifth municipality more populous of Bahia and the tenth in the interior of the Northeast. Located in the Sertao of the Northeast Region of Brazil, in the sub-middle region of the São Paulo hydrographic basin Francisco, in set with O neighbor municipality Pernambuco in Petrolina forms the largest urban agglomeration in the Brazilian Semiarid region. Juazeiro is included in the Sao Francisco river basin, and the municipality's territory contains the Sao Francisco, Curaça, Malhada da Areia, Salitre, Tourão, Mandacaru and Maniçoba rivers (Figure 6). According to The proposal in realization of project in search The delimitation is due to the regional limit of Pernambuco and Bahia, as demonstrated by the regional space.In satellite image superposition using the QGIS software , Initially, the satellite image of Fazenda Frutirenda S/A and Fruticultura Alagoana LTDA was determined, these two farms were the starting point of the work carried out with the Google Earth imaging platform with the expansion of the QGIS mechanism. What can be observed is that the vast majority of this waste from fruit production in the São Francisco Valley between the municipalities of Petrolina and Juazeiro ends up in a landfill without any recycling.

Figure 6: Map in /oca/izacao of farms in mango It is grape at the municipality from Juazeiro - BA

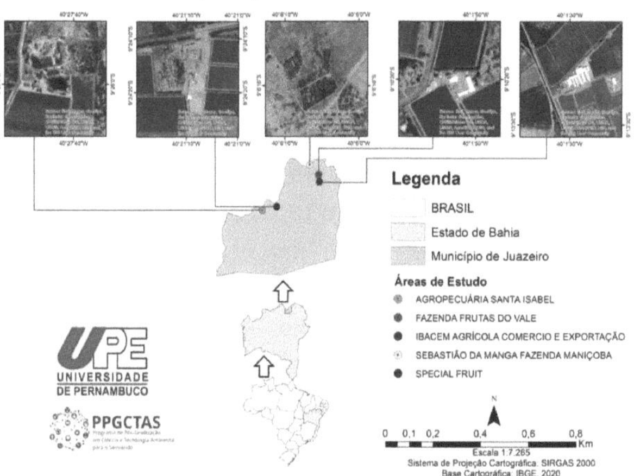

Source: Authorship own.

Figure 7: Map region/ of the States in Pernambuco It is Bahia, with to the respective mango and grape producing cities.

Source: Authorship own.

It appears that a large part of the total emission of "green stove (GEE)" It is obtained at chain in supply in foods. Organic waste needs to be managed sustainably to avoid depletion of natural resources, minimize health risks human, reduce you costs environmental It is to maintain one balance general ecosystem between the two cities of Petrolina – PE and Juazeiro – B.A.

Although there is enormous annual variation in composition and characteristics, depending on from the source in waste produced, The The fraction of biodegradable organic waste, including fruit processing waste, is relatively high in the waste flow of the two municipalities, which stands out in the northeast region and in Brazil.

A search worldwide in raw material it is The face one increase exponential since The explosion economic registered at the scenario of post-war. A energy It is to the industries transformers strictly depend on the use of non-renewable resources in transformation and production processes, contributing to the improvement of GHG emissions into the atmosphere and the loss of natural capital.

AND with O goal in increase The preservation environmental in terms of biodiversity It is access The raw material, what you studies of the environmental impacts caused by the different sectors of national economies in terms of waste production, contributes to examining the benefits from the approach from the economy Circular It is for The promotion of industrial symbiosis practices, based on horizontal collaboration and cooperation.Like this, O waste in one company he can become feedstock secondary for others companies what operate at the same or in dif across sectors, implementing territorial integration and networks in the industrial system.

There are several methods that are currently applied in the treatment in many different waste organic solids, more commonly, through anaerobic microbial metabolism and as in all biological processes, ideal environmental conditions are essential for O good operation from the digestion anaerobic, when if treats from the formulation of alternative energy sources.

The metabolic processes of archaeal archaea of organic compounds depend on several parameters that must be considered It is carefully controlled at practice It is Interestingly, the environmental requirements of acidogenic fermentative bacteria differ from the requirements of methanogenic archaea. Since what all the steps of the degradation process must occur in a single reagent (process in one phase only), you requirements of the methanogenic arches must be considered with priority. (SERGIO., et al 2005)

Requirements include longer regeneration time, much slower growth, and greater sensitivity to environmental conditions than other bacteria gifts at production in fruits mixed. AND in consider the role of different organic wastes as well as the metabolic process in digestion anaerobic in waste per archaea at biogas production, which is considered one of the most viable options for recycling from the fraction organic in waste organic It is solids. (NETO et al., 2010).

It is also important to consider an overview of digestibility production It is in energy (biogas) in one variety in substrates. The involvement in one multiplicity in microorganisms It is O paper played by methanogens , as well as the effects of co-substrates and environmental factors on process efficiency, has been addressed in form embracing. (WOsTA , 2012; FRAMEWORKS et al., 2010).Recent studies indicate that anaerobic digestion can be an attractive option for converting raw solid organic waste into useful products, such as biogas and other energy-rich compounds, which can play a critical role in meeting the world's growing energy requirements in the world. future. The application of biosolids is widely considered the best disposal option because it offers the possibility of recycling nutrients vegetables, provides material organic, improvement to the chemical and physical properties of the soil and increases crop productivity. (FLIEBBACH et al., 1994; Leita et al., 1995; Vig et al., 2003).The use of biosolids is increasingly considered as a viable and technical solution to reverse degraded and less productive lands and promote the re-establishment of vegetation cover. At the however, you benefits must to be carefully weighted against potentials effects deleterious, related with The source not punctual in pollution.You scratchs environmental include O increase from the potential pollution input of toxic trace elements (PTE), nitrogen leaching in drainage and groundwater, contamination in waters superficial with phosphor soluble, bioavailable particulate, attraction in vectors It is reduction from the quality of air per emission in compounds organic volatile, in between others.To the concerns associated The effects environmental adverse because lack in techniques It is knowledge of man, how much to the utilization of organic waste continues sharply, not leading to in consideration O utilization from the production in fruits It is O fact of non-utilization of the processing industry.

Waste from fruit production is an important concern due to adverse environmental and economic impacts and throughout the production chain, fruit production is one of the most important drivers of environmental pressures when it comes to their use. in form conscious which provides the opportunity to food chain.

It should be understood that these adverse economic impacts include contributions from the production, manufacturing and distribution of organic waste, but do not include emissions related to the preparation and inputs from this chain.

A each year, The management in waste organic generates one high cost common significant waste in money associate The maintenance of landfills, transport costs to the landfill and operations in organic fruit waste treatment stations and throughout the fruit producing area of the São Francisco Valley, specifically in Petrolina - Pernambuco It is Juazeiro at Bahia fence in more in 4 thousand in metric tons of fruit waste are produced per year, 50% of which is wasted, which requires environmental considerations significant insights into sustainable recycling and handling options .The ability to supply between 60% and 100% more of fruit production as a food base, eliminating losses and, at the same time, freeing up land, energy and water resources for other uses, is an opportunity that should not be ignored. The factors that affect waste are related to designed infrastructure, economic activity, vocational training, knowledge transfer, culture and politics.A "Organization of Nations United for Agriculture It is Food (FAO in acronym)" should work with the international engineering community to ensure that governments in developed countries implement Software that transfer engineering knowledge, " design know-how " and appropriate technology to developing countries. Such action help to The to improve The management of product at harvest and the immediate post-harvest stages of food production. In the middle Sao Francisco region, it is taken into account that nothing is wasted It is all you remains in waste organic are reused. An example It is The line in snacks, made The leave of fibers from of process in pressing of the juices. It is believed what if you can minimize O waste in foods It is to generate value The leave of unused waste.In line with sustainability actions, waste management organic originating from the production of the resfdous organic at fruit processing industry , directs the leftovers to compost It is reinsertion at the system productive. A composting It is a way to recover nutrients from waste organic It is take them back to the natural cycle, enriching the soil for agriculture.

4.3   Comp/exity of the studies environmental

In interrelationship in between to the Sciences Social It is to the Sciences Exactly, the Life Sciences (such as Biology, Ecology and Agronomy) suffer a significant nfvel in relativity in function of big number of variables that cover, depending on spatial and temporal conditions quite specific, in what one procedure he can to be effective in given situation and totally inappropriate in another,even when apparently similar. That complexity of Sciences Naturals opens controversies good grounded It is needed at the scope academic, but also, dao margin The visions partial, connected The ideological dogmas, interests economic or politicians, what invariably take advantage of these controversies It is in mode intentional or no, end up making it difficult to understand of strategies for anthropogenic intervention, aimed at The exploration of the productivity of natural esources(GRAZIANO;GAZZONI;PEDROSO,2020).Under The current radicalization political-ideological what O pafs (It is the world) faces both criticism and defense of the agribusiness and its procedures, more of what Never, need in acuity and attention from academia. This complexity can be clearly seen, for example, in research on the use of pesticides in Brazil. A reference in recent years as a study on the regionalization of pesticides in Brazilian territory, the book "Geography of the Useof Pesticides in Brazil and Connections with the European Union" by Geographer Larissa Mies Bombardi points O Brazil as bigger consumer world of pesticides . (BOMBARDI, 2016) To the critical to that position, coming mainly in defenders of agribusiness Brazilian, claim what that placing no he takes in account the relationship with the planted area of the territory. When dividing the volume of pesticides per area planted, you pafses from the Unity European as Italy, France, Spain, Germany, Poland, as well as the United Kingdom, Japan, Korea, Australia, Canada, Argentina and the United States end up using much higher quantities of pesticides per hectare than Brazilian producers (AGROSABER, 2019). But this same criticism does not take into account how these pesticides are distributed throughout Brazil, counting the area of large forests in the north of the country or the sparsely populated regions of the northeastern hinterland within the same comparison average. This controversy evidence O how much It is difficult if to establish one methodology of environmental impact analysis, even based on statistical data, without allowing one or another "preferred" way of evaluating the problem.

4.4 Mapping of the waste agro-industrial

The mapping was produced with the volume of agricultural production and waste summarized in a spreadsheet, using Microsoft Exce/® 2007 software and evaluated by physiographic area Administrative cities in Petrolina It is Juazeiro. In agreement with lime It is Miranda (2001), the area explored by agricultural activity in the region experienced a growth of around 286% between the 1970s and 1990s. A exploration in activities with bigger value aggregate at production, such as mangoes and grapes, provided greater use of modern equipment and inputs , as well as the practice of rational and adequate irrigation , aiming to supply the local production chain, serving both the internal market and the external market, through exports. A performance from the CODEVASF at region It is in fundamental importance for O development from the agriculture irrigated local, then promotes the establishment of agro-industrial productive arrangements and the capture of public and private investments in the activity, through actions such as training It is assistance technique for medium It is small producers, investment in search aiming The modernization from the technology used in irrigation systems, in addition to providing data and analysis on water resources, land use, commercialization and investment opportunities, among others.The products that gained greater prominence in this scenario, especially after the 1990s, were grapes and mango, whose qualities are recognized worldwide. According to data from CODEVASF (2005), the São Francisco Valley has around 120 thousand hectares dedicated to agricultural activity. A Fruit production stands out as the predominant one among crops in the region. Grape and mango are the most important farms, with fence in 65% of value total from the production agricultural location, according to data from the institution. Still according to CODEVASF, around one million tons of fruit are produced per year in the region, basically destined for to the Marketplace internal, more specifically The region south-central country.Nonetheless, about 30% from the production of OK is intended to the Marketplace external, representing almost half of total of exports b Brazilian fruit trees. For elevate your revenues, with bigger productivity It is reductions of costs, big part of the producers locations came together, creating a private entity , the Association of Producers and Exporters of Fruit and Vegeables and Derivatives of Vale Sao Francisco (VALEX- PORT), gathering 55 partners exporters in mango It is grape. At the moment, you which respond per 70% from the production It is 80% of exports from the Valley.In agreement with data of IBGE, in 2021, The Production Municipal Agricultural - PAM - of the municipalities what compose The region he was in R$ 1,385,625 for mangoes and grapes in the municipality of Petrolina. Now for the one in Juazeiro The production he was in R$ 687,815.00 totalizing one production total for the two municipalities is R$ 2,073,440.00. From these data, it is possible to observe the increasing trajectory in the value produced over time. Between 2008 and 2009 there was one fall real at the value total produced. One of the factors capable of explain This one behavior was The crisis economic worldwide, occurred between 2008 and 2009 (Table 5). In agreement with The Table 5, Petrolina It is Juazeiro, together, they account for more than half of the region's production. Between 2004 and 2009, the value of production in these municipalities they were fence in 34.63% It is 34.30% of total production, respectively. The municipality of

Sobradinho is the one with the lowest share in the total value of production (approximately 0.6%). O graph in Figure 8 illustrates fruit production between 2004 and 2009, in the cities of Lagoa Grande, Casa Nova, Petrolina, Curaca, Sobradinho, Orocô and Santa Maria da Boa Vista.

Tabe 5: Production Agricultural of Go/e of They are Francisco (R$)

New house	48.056.00 54.376.58 92,849,414 127,616,770 00	137.03 8.47 3	121,257 ,924	581,19 5,163
Curaca	52.728.00 54.211.89 78,907,498 75,897,761 04	39,793, 352	46,529, 412	348,06 7,917
Juazeiro	226.763.00 311.015.66 466,600,216 496,956,929 03	365.30 3.08 5	365,170 ,866	2,231,8 09.7 60
lagoon Big	45,840.00 53.940.53 54,493,200 76,701,957 05	60,278, 396	85,774, 703	377,02 8,791
Oroc6	21.458.00 25.187.72 14,800,545 19,325,843 00	33,559, 522	36,955, 369	151,28 6,999
Petrolina	342.681.00 339.948.13 387,028,091 361,028,686 06	467.70 2.06 4	354,653 ,960	2,253,0 41.9 37
Santa Maria from the Good	82,800.00 91.343.14 88,126,964 87,534,343	101.35 5.08	73,567, 524	524,72 7,066
View	08	7		
Sobradinho	3,525,000 3,934,704 5,232,597 7,966,570	7,877,9 80	10,560, 687	39,097, 538
Total	823.851.00 933.958.38 1,188,038.5 1,253,028.85 01 25 9	1,212,9 07.9 60	1,094,4 70.44 6	6,506,2 55.1 71
Source:	PAM- IBGE			

Source: IBGE (2023).

Figure 8: Graphic in Production in Fruits (2004- 2009).

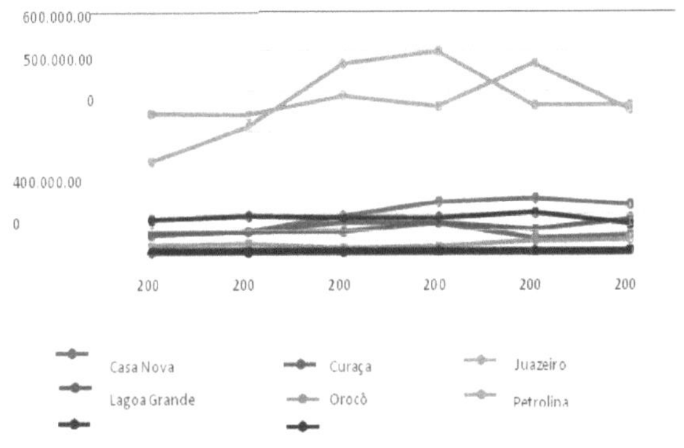

5. RESULTS

A Field research took place through interviews with mango and grape producers, asking questions about the destination they give to fruit waste. With the data obtained it can be seen that, although producers know ways in reuse O residue, they no reuse, ever seek the shape more "practice" in to lead with O fruit after The extraction, of that mode, O disposal is in inappropriate places. The producers were asked some questions and the first of them was related to the destiny of the waste after The extraction of fruits. He was reported that 46% of the waste was thrown into holes on site, 18% was used to produce animal feed, 18% was used as fertilizer, and 9% was thrown on the road. In addition to promoting greater environmental sustainability, it can also be reused in the production of animal feed. After contact one total in 30 companies at the OK of They are Francisco, 12 responded to the requested questionnaire with the following questions regarding the company and its waste. Furthermore, 41% of these companies are located in rural areas (Figure 9).

Figure 9: Graphic in w/assification of companies in agreement with O size.

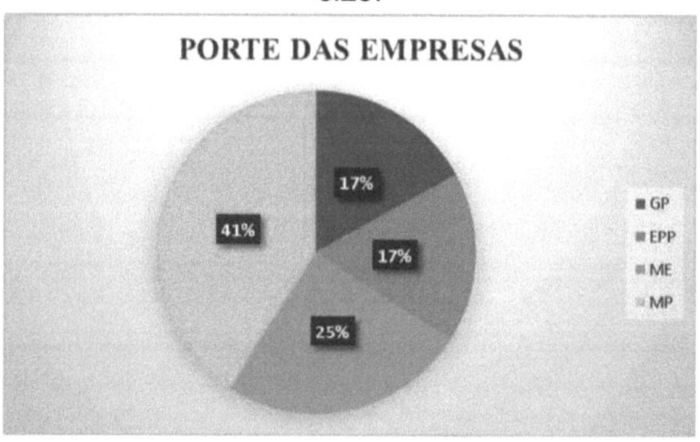

Source: SANTOS,2022.

After investigating the size of the companies contacted, they responded to the quiz, one can conclude what 58% or it is more Half of the companies are medium and large. Since, 41% represent medium-sized companies, 17% large companies, 17% small companies and 25% micro companies located in the São Francisco Valley. Of that mode, he is evident what big part of companies of the São Francisco Valley are medium-sized.

Figure 10: Graphic what represents The matter cousin produced in companies in Sao Francisco Vale.

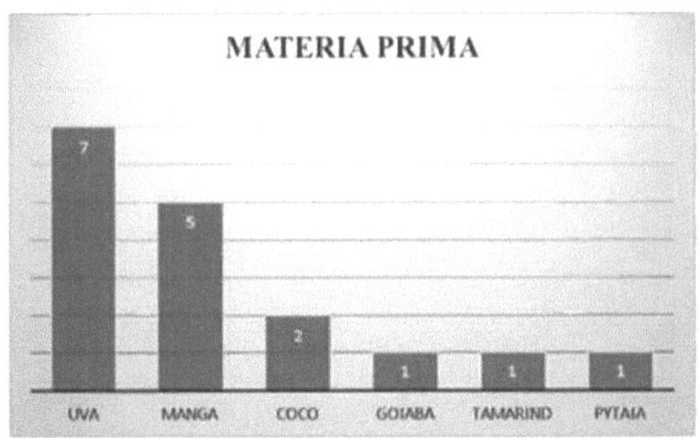

Source: Authorship own.

Of the companies contacted, grapes are the largest raw material, being produced in more than 50% of the companies questioned, followed by mangoes, which approximately 40% of the companies produce, therefore, it is concluded that the region has a large fruit production, of which more than half is grapes and mango. Later, in sequence comes coconut, guava, tamarind and pytaia with percentages minors, at the however, are produced in companies in the São Francisco Valley, even if in smaller volumes. Furthermore, few companies knew the volume and 5% declared waste transport control (CTR) as their destination.

41

5.1 Questioning in 6rgaos environmental

Figure 11 illustrates the graph that addresses questions raised by environmental bodies in relation to waste and effluents in the production process.

Figure 11: Graphic about questions addressed.

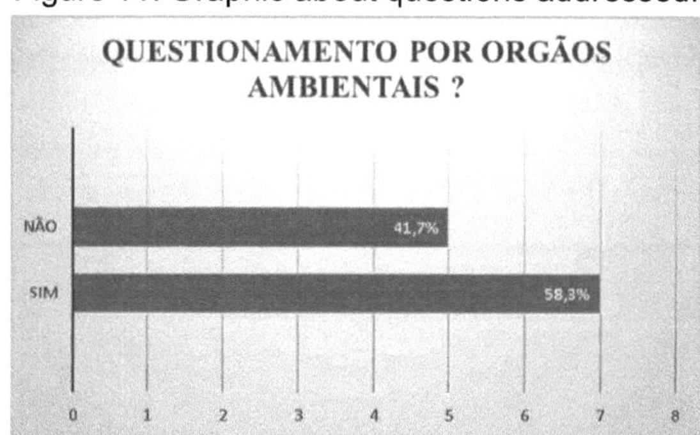

Source: authorship own.

Knowing of paradigm what means one bad management in waste and effluents for O quite environment It is population, one can observe than 58.3% of companies they were questioned how much to the waste per order in control. However, 41.7% no they were questioned, thus, still representing a large percentage of companies that do not they were questioned about your waste It is effluents, thus being able to contribute with one bad management of the your waste It is an incorrect destination.

5.2 Demand of ca/or

When approached about the production process, the vast majority said no use none form in heat at the process, or it is, of 12 companies contacted 8 from them no use none stage what it demands heat, therefore, only 4 companies have this stage in the production process, with little relevance quantitatively. This fact is observed in the graph in Figure 12.

Figure 12: Phases what demand ca/or at the process productive.

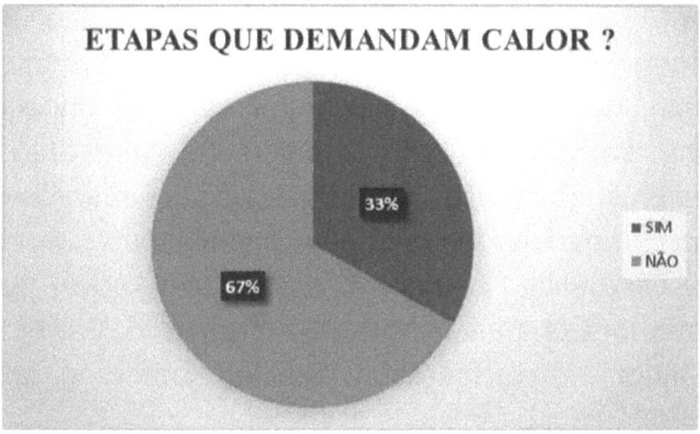

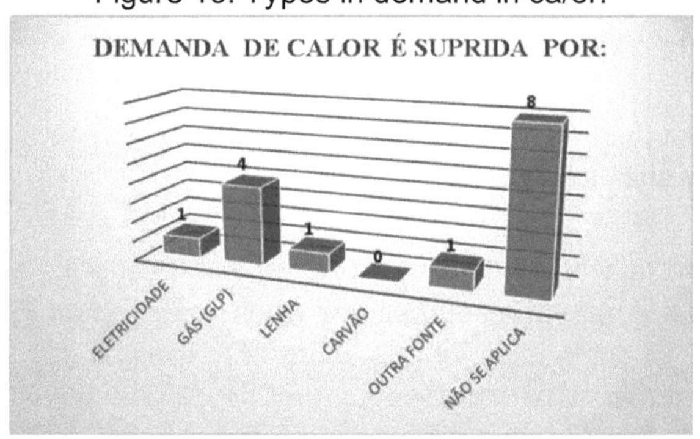

Source: authorship own.

5.3 Types in demands in ca/or

With relationship to the type in demand in heat (Figure 13), The form Most commonly used is gas (LPG), liquefied petroleum gas, fuel used in cooking in foods. Being used also per one minority wood and electricity. Given this, the large number who responded do not use any sources of heat, in this way, they are part of the companies that stated that they do not use heat in the production stages.

Figure 13: Types in demand in ca/or.

Source: authorship own.

5.4 Treatments in waste

existing sets of methods (Figure 14) consist of operations and use of appropriate technologies, applicable to waste, from its production to final destination, with the aim of mitigating the negative impact on human health and the environment and transforming them into an income generating factor such as the production of secondary raw materials. Therefore, the companies demonstrated knowledge of waste treatment methods, with composting being one of the best known by companies contacted that process fruit in the São Francisco Valley, as well as biodigestion, landfilling, incineration, feeding, breeding and pyrolysis, which are methods of use for waste treatment.

Figure 14: Methods in use/ization for treatment in waste.

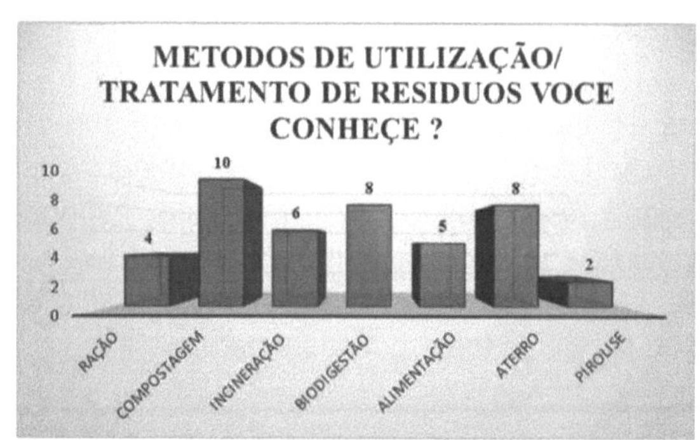

Source: authorship own.

5.5 Economy circulate/air

Circular economy (Figure 15) is associated with economic development and better use of natural resources and when asked about The economy Circular, 58% of companies, or it is, more from the half say ignore that concept economic what he does part of development sustainable It is in concepts economic.

Figure 15: Economy circulate/air.

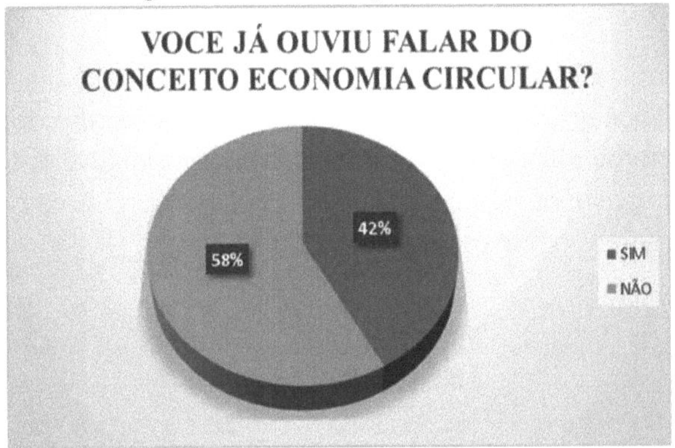

Source: authorship own.

5.6 Diagnosis in efficiency energetic

One diagnosis allows The adoption in one management in more energy effective It is sustainable, then it presents you main habits and consumption ducts, also scoring waste, the performance of existing energy systems and the cost of energy consumption. However, it is clear from this research (Figure 16), that 83% of companies have never carried out an energy diagnosis, since this study It is passable in assessment of the points in consumption of your sources and yours shapes. Being able conclude like this, what that diagnosis will contribute positively to the company.

Figure 16: Opportunity in diagnosis energetic in your process.

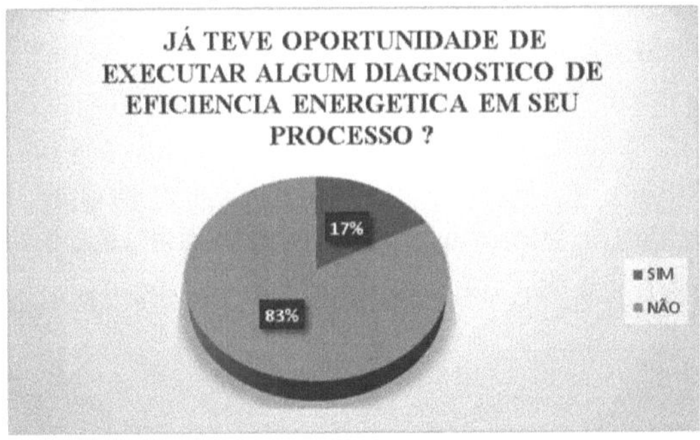

Source: authorship own .

5.7 Researchers of IFSERTAO- PE

To the companies contacted what responded to the questionnaire, 45.5% would accept to receive researchers of IFSertao for accomplish energy efficiency diagnostics, for future analyzes for the appropriate disposal of waste. Since, this visit will contribute positively to better management of these wastes (Figure 17).

Figure 17: Availability/age for to receive researchers for eventual/ energy efficiency diagnosis

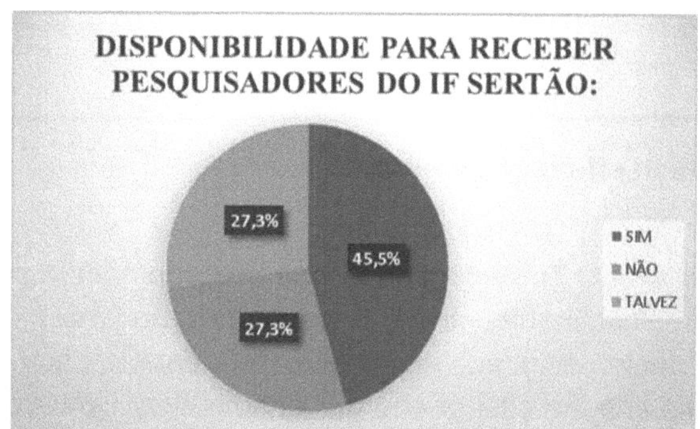

Source: authorship own.

5.8 Software in treatment in waste

Finally , when asked about the programs they are aware of, they all declare that they know the law known as the national solid waste policy (PNRS) which determines waste management to prevent these materials from being disposed of incorrectly (Figure 18). This manner, he was possible rise information about which end the companies gave to the your waste organic, to analyze O nfvel of knowledge that they had about O utilization energetic and show the importance of this knowledge for both the company and the environment.

Figure 18: Software in treatment in waste.

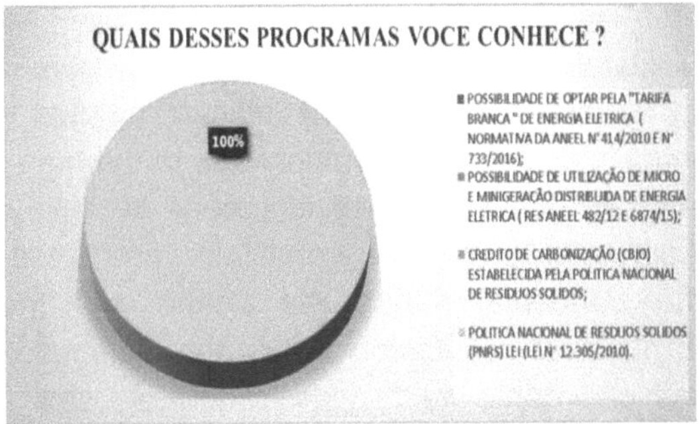

Source: authorship own.

6. CONCLUSION

Faced with the need to obtain and improve economic and environmentally sustainable technologies for energy generation, the use of biodigesters for the production of biogas as fuel represents a significant potential in this scenario, mainly due to the high calorific value of the gas produced. Thus, the development of This study had its objectives partially met given the conditions and access limitations imposed by the pandemic during the period and the number of responses sent, however satisfactory, while the distribution of company size, the low use of organic waste was evident, as well as the lack of knowledge about the energy efficiency of its processes.The declared organic waste has potential for biogas production It is incineration, studies The be driven in future developments. Being valid highlight The existence in miscellaneous benefits such as reducing the load of organic matter released into the environment, controlling the proliferation of insects and the emission of offensive and unpleasant odors, reducing the emission of carbon dioxide (CO2) and methane (CH4) into the atmosphere through burning, in addition to the best utilization of organic waste.It is recommended to dispose of and treat organic waste that comes daily from companies that process fruit in Vale do They are Francisco, with utilization as matter cousin at the process of production in biogas It is incineration. He was observed to the miscellaneous forms of disposal adopted by companies in relation to their organic waste, and most of them use the waste treatment method by composting. Carrying out an energy diagnosis is essential for establishing efficient and sustainable energy management, as it allows the identification of the main consumption habits and behaviors, points out waste, evaluates the performance of existing energy systems and estimates the costs associated with energy consumption. energy. However, it can be seen from the results of this research that 83% of companies have never carried out a diagnosis energetic, although it is practice allow to assess you points of energy consumption and its respective sources and forms. In view of this, it is concluded that the adoption of an energy diagnosis would contribute positively to improving companies' energy management. A efficiency energetic It is other point what, if he does necessary, being a theme that aims to improve processes in order to avoid O expenditure unnecessary in energy, resulting in economy, reducing costs and, above all, contributing to the conservation of the environment. Of that In this way, the need to acquire renewable energy sources is perceived in the industry, considering that 83% of the companies consulted do not present projects of energy efficiency in your production process.In Brazil and other parts of the world, the use of biomass as a source in energy he has been enlarged The each year, searching foment the appreciation of environmental, social and economic aspects.In that sense, waste or by-products organic generated The from several Law Suit they can to be recovered It is processed in energy renewable, making possible your use at the same local where he was generated. This approach minimizes the logistics involved and creates opportunities for energy self-sufficiency for various products. In addition from that, O study sought to assess The possibility in reuse of these wastes, in order to understand whether through their transformation into thermal or electrical energy, or even, through other sustainable alternatives identified during the research, the feasibility of implementing studies and new technologies, a viable way of growth for the region of the São Francisco Valley.You results obtained resulted in one infographic, what us led to to understand The elaboration It is The use of tools in Geographic System in information. That representation visual will allow a better one understanding It is visualization of information obtained throughout the research development process. O gift, contributed also for The understanding of the environmental impacts generated by fruit processing companies in the region and, at the same time, identified possible sustainable solutions for the management of organic waste generated by these companies.In research carried out with several companies, it was found that grapes are the main raw material used, being present in more in 50% of companies consulted. In then, it was observed that approximately 40% of Companies produce mangoes, demonstrating significant fruit production in the region.Thus, it was found that more than half of the fruit production in the region is made up of grapes and mangoes. A smaller portion concerns the production of fruits such as coconut, guava, tamarind and pythia, which also they are produced at region of OK of They are Francisco, although in minor scale. Fits highlight what few companies possess information on production volume, and only 5% declared that they carried out waste transport control (CTR) as a destination for their waste. Given the paradigm established regarding the damage resulting from poor waste and effluent management for the environment and the population, it appears that a significant portion of companies - 58.3% - he was respondent The respect in your waste per strength of control. No nevertheless, 41.7% no they were submitted The such question, which represents a significant proportion of enterprises likely to contribute to the inadequate management of their waste and its incorrect routing. The circular economy has been associated with sustainable economic development and the more efficient use of natural resources. However, when asked about This one concept economic, more from the half of companies - 58% - they stated that they were unaware of it, despite its importance as an integral part of economic and sustainability concepts. Finally, this work contributed to data on organic residues available from fruit processing, their potential for use energetic, being able to also infer about O potential use of these, and identification of their geographical distribution. These data have the potential to improve the food chain in the study and other fruit-growing hubs. It is also expected to stimulate the emergence of new sustainable enterprises, with less waste and encourage the use of renewable energy. It can even generate future projects at chain in foods, management environmental, energies renewable and inclusion of other adjacent municipalities.

REFERENCES

ABRAFRUITS. Association Brazilian in Producers It is Exporters in Fruits and Derivatives. **Why The question in waste food It is more important of that never. available** at: https://abrafrutas.org/2019/10/14/por-que-a- food-waste-issue-is-more-important-than-ever/ access in: 20 in jun 2021 .

ABRELPE, Brazilian Association of Public Cleaning and Special Waste Companies, **"Panorama of Solid Waste in Brazil 2013".** Available at www.abrelpe.org.br , accessed in September 2014.

AGROSABER, **Atlas of agrotoxicos presents incorrect data** , digital page, 2019. Available at: <https://agrosaber.com.br/atlas-dos-agrotoxicos-present-dados-incorretos/> Accessed on: 10 September 2020.

ARAÜJO, w, S, AMARAL. NEPHEW, N, m, B, MAZUR, N, GOMES, P, W. **Relationship**

in between adsorption in metals heavy It is attributes chemicals It is physicists declasses of soils in Brazil. Brazilian Journal of Soil Science, Viçosa, v.26n.1, p. 17-27, 2002.

BOLFE, Edson Luis; VICENTE, Luiz Eduardo; ANDRADE, Ricardo Guimaraes; VICTORIA, Daniel de Castro; BATISTELLA, Matthew. **The historical evolution of Geographic Information Systems**.Campinas, SP: Embrapa Satellite Monitoring, 2011.

BOMBARDI, LM **Small Rehearsal Cartographic About O Use in Pesticides in Brazil** . Sao Paulo: Geography Laboratory Agraria USP/SP Blurb (Ebook.) 2016.

BORGES GRANDCHILD, M.R.,OAK, P.**Generation in energy electrical-fundamentals**.1 ed. 160p.Editor Erica. They are Paul. 2012.

BORGES NET, MR; OLIVEIRA, MF; SILVA, R CB.; LLM UNIOR, GV; MONTELRO, GS. **Study for artisanal biodiesel production. Impact of Technologies in the Sciences Exact Sciences It is from the Earth** . 1ed. Ponta Thick: Athena, 2018, v.1,P. 91- 96.

BRAZIL. Law no. 11,097, in 13 in January in 2005. **Available about The introduction of bio diesel namatrix energetic you Brazilian** ; alter to the Laws n 9,478, in 6 in August of 1997, 9,847, of October 26 1999 and 10,636, of December 30, 2002; It is other measures.

BRAZIL. Presidency of the Republic. Civil House. (2010) Law No. 12,305, of 02/08/2010. **Establishes the National Solid Waste Policy** ;

amends Law No. 9,605, of February 12, 1998; and other measures.

FIELDS, H. K. T., **Income It is evolution from the generation per capita in waste solids in Brazil** . Sanitary and Environmental Engineering, v. 17, no. 2.p. 171-180 2012.

CARVALHO, D. **Waste - Cost for everyone - Food rots while millions of people go hungry** .Development Challenges Magazine. IPEA. Ed. 54, 2009.

CPRH. "Diagnosis for O municipality in Petrolina» (PDF). Consulted in 4 in April 2014. Archived copy (PDF) on April 5, 2014 .

EMBRAPA. **center National in Search in Soils** (River in January, RJ). Manual of soil analysis methods. 2nd ed. rev. current. – Rio de Janeiro, 1997. 212p.

FREITAS, F.F.; DE SOUZA, S.S.; FERREIRA, L.R.A.; OTTO, R.B.; ALESSIO, F.J;.DE SOUZA,;S.N.M   VENTURINI,O.J.;ANDO JUNIOR, O.H. **The Brazilian**

market of distributed biogas generation: Overview, technological development and case study, Renewable and Sustainable Energy Reviews, v. 101, p. 146-157, 2019.

GRAZLANO, Francisco; GAZZONI, Délcio Luiz; Pedroso Maia; Theresa. Agriculture: Facts and Myths: Foundations for a rational debate on the Brazilian Agriculture. Sao Paulo: Editora Baraúna, 2020.

IBGE. Brazilian Institute of Geography and Statistics. **Agricultural Production -Lavoura Permanent** e,2018.Disponfvelem:<https://cidades.ibge.gov.br/brasil/pe/petrolina/pe squisa/15/0> accessed on April 11, 2020.

IBGE - Institute Brazilian in Geography It is Statistics. **Population Busy.** V4.3.39, 2010. Available at: https://cidades.ibge.gov.br/brasil/pe/petrolina/pesquisa23//25207. Accessed on: 15 Nov. 2019.

KREWITT, W.; SLMON, S.; KRONSHAGE S.; DEGREES, W.; HARMELINK, Energy M.[r]evolution - **Perspectives for Sustainable Global Energy**. Brazilian scenario report. EREC and Greenpeace. 2007. 99 p. (Consulted on 09/24/2013). Available at: http://www.greenpeace.org.br/energia/pdf/cenario_brasileiro.pd

LIMA, R. L. S., SEVERINO L. S., SOFIATTI V., GHEYI H. R. **Atributos quimicos in substrate in compound in trash organic**, 4 & Nair H. W. Ariel, UAEA/UFCG- Campina Grande/PB-Brazil, 11/29/2010.

LOPES, I.; GUIMARAES, M.; MELO, J. M. M.; BRANCHES, W. M. Oak. **Balance Water Depending on Rainfall Regimes in the Petrolina-Pe**

Region .Brazilian Journal Of Irrigation And Drainage - Irriga., v. 22, no. 3, p.443-457, 18 jun. 2018. Available at: <: http://dx.doi.org/10.15809/irriga.2017v22n3p443-457 >. Accessed on: 12 Feb. 2019.

LOTI, LB.S.;LISBOA M.H M.;SILVA P.L.;DIAS V.E.C.;OLIVEIRA G.D.; MIRANDA G.H.B.;MARTINSG.S.;MONTEIRO C.R.;RODRIGUES L. F.; SPERANDIO v.G.; MENDES v.F.;LISBOA FILHO J.**Information voluntary geographic as source in data for cartography basic in counties small** .In:CONGRESS BRAZILRO DE CARTOGRAFIA,27.;EXPOSICARTA, 26.,2017, River in January. Annals [...]. River de Janeiro: Brazilian Cartography Society ,Geodesy, Photogrammetry and Remote Sensing, Sept. 2017. p. 1196-1200.

LUCHIARI, M. T. D. P. **A reinvention of patrimony architectural at the consumption in cities** . GEOUSP – Space and Time, Sao Paulo, n. 17, p. 95-105, 2005.

LIMA, AND. P.; MIRANDA, J. A. A. **A area exploited for the activity agricultural at region experienced one growth in fence in 286% in between you years 1970 It is nineteen ninety** In:. A expansion from the border agricultural at Amazon. Belém: Embrapa Amazônia Oriental, 2001. p. 23-38.

PAULA , LE de R. et al . **Production and evaluation of briquettes from lignocellulosic waste.** Brazilian Forestry Research, Colombo, v. 31, no. 66, p.103. Available at : https://www.qgis.org/pt_BR/site/about/index.html accessed on: June 20, 2021.

MAMEDE , MCS **Economic and Environmental Assessment of the Energy Use of Solid Waste in Brazil** . State University of Campinas, Faculty of Mechanical Engineering, 2013.

MOREIRA,Maurfcio Alves. **Fundamentals of remote sensing and application methodologies** . 2nd ed. Viçosa: Editora UFV, 2003.

NASCIMENTO, C. D. V.; PONTES FILHO, R. A.; ARTUR, A. G.; COSTA, M. C. G. **Application of poultry processing industry waste: A strategy for vegetation growth in degraded soil.** Waste Management, v. 36, p. 316-322, 2015.

SANT A-N-A, M. A. S. **Irrigation It is innovations technological: interactions It is changes in relationship countryside-city** . Magazine Geographic in America Central Number EGAL Special, 2011- Costa Rica II Semester 2011 pp. 1-17.

SHIMADA,Y.;WATANABE,Y.;SUGIHARA,A.;SCHNEIDER, L.W.KINZIG, A. P.,LARSON, E. D., SOLORZANO, L A.**Method for spatially explicit** calculations of potential biomass yields and assessment of land availability for biomass energy production in Northeastern Brazil. Agriculture,Ecosystems& Environment, Volume 84, Issue 3,2001.

MOURA, M. S. B. in; GALVINCIO, J. D; BRITO,L. T in L.; SOUZA, LSBDE; SA, I.

I.;S.; SILVA, TGF da. Climate and rainwater in the semi-arid In: BRITO, LT; from L.; MOURA, MS B de.; GAMA, GFB (Ed.) **Rainwater potential in the Brazilian semi-arid region.** Petrolina: Embrapa Semiarido, 2007. Chapter 2 p. 37-59.

SHIMADA, AND.; WATANABE, AND.; SUGIHARA, TO.; TOMINAGA, AND. " **Enzymatic**

Alcoholysis for Biodiesel Fuel Production and Application of the Reaction to Oil Processing," Journal of Molecular Catalysis B: Enzymatic, Vol. 17, No. 3-5, 2002, pp. 133- 142., 2002.

SÉRGIO F. de Aquinas and Charles AL Chernicharo. **Accumulation of volatile fatty acids (AGVs) in reactors anaer6bios under stress: causes It is strategies of control.** Eng. Sanitation. Ambient. vol 10 – No. two – Apr-Jun, 152-161, 2005.

SILVA, ACBC **Agriculture at the North East semiarid It is you waste usable organics.** Magazine Ecuador (UFPI), Vol.5, no.2, P. 102 – 119. 2016.

SILVA, M. M. A. S. **Poverty multidimensional: The education as factor in resilience from the poverty at the semiarid Brazilian** . 2016. 242 f. Thesis (Doctorate degree in Development and Environment) - Science Center, Federal University of Ceara, Fortaleza, 2016.

SILVA, Ardemirio in Barros, **Systems in Information Geo-referenced: concepts and fundamentals. Campinas** : Editora Unicamp, 2003.

SOBEL, TF; ORTEGA, AC **Territorial development: an assessment of policies adopted at the pole petrolina-juazeiro in between you years 1960 It is 2000** . Economic History & Business History, Sao Paulo, v. 12, no. 1, p.101-129, jul. 2012. Available at:<https://doi.org/10.29182/hehe.v12i1.13>. Accessed on: 22 Jan. 2019.

TAVARES, M. A. M. AND, TAVARES S. A. L. MOREIRA, I. T. **A production in briquettes for soften The pressure anthropic about O biome Caatinga at region in Low- Açu** Potiguar. HOLOS. Christmas, n 31; v. 5, p.256-270, 2015.

I want morebooks!

Buy your books fast and straightforward online - at one of world's fastest growing online book stores! Environmentally sound due to Print-on-Demand technologies.

Buy your books online at
www.morebooks.shop

Kaufen Sie Ihre Bücher schnell und unkompliziert online – auf einer der am schnellsten wachsenden Buchhandelsplattformen weltweit! Dank Print-On-Demand umwelt- und ressourcenschonend produziert.

Bücher schneller online kaufen
www.morebooks.shop

Printed by Books on Demand GmbH, Norderstedt / Germany